ULTIMATE SUPERCARS

FERRARI F8 TRIBUTO

By Ellen Labrecque

Kaleidoscope
Minneapolis, MN

The Quest for Discovery Never Ends

....

This edition first published in 2021 by Kaleidoscope Publishing, Inc.

No part of this publication may be reproduced in whole or in part without written permission of the publisher.

For information regarding permission, write to
Kaleidoscope Publishing, Inc.
6012 Blue Circle Drive
Minnetonka, MN 55343

Library of Congress Control Number
2020936263

ISBN
978-1-64519-262-6 (library bound)
978-1-64519-330-2 (ebook)

Text copyright © 2021 by Kaleidoscope Publishing, Inc. All-Star Sports, Bigfoot Books, and associated logos are trademarks and/or registered trademarks of Kaleidoscope Publishing, Inc.

Printed in the United States of America.

Bigfoot lurks within one of the images in this book. It's up to you to find him!

TABLE OF CONTENTS

Chapter 1: *Bellissimo!* **4**

Chapter 2: *Built For Speed* **10**

Chapter 3: *The F8 Tributo's Best Features* **16**

Chapter 4: *The Tributo's Future* **22**

Beyond the Book *28*

Research Ninja *29*

Further Resources *30*

Glossary *31*

Index *32*

Photo Credits *32*

About the Author *32*

Chapter 1
Bellissimo!

Sam Harris is cruising in the beautiful countryside of Italy. His dad is driving. They are in a bright red Ferrari F8 Tributo. The F8 Tributo is a supercar. A supercar is a car that is fast like a race car.

FUN FACT
The F8 Tributo can reach a blistering speed of 211 miles (339 km) per hour!

But it is legal to drive on regular streets. Sam sits in the passenger side of this two-seater car.

The F8 Tributo hugs a curvy road into a village. Kids are walking home from school. They jump up and down in excitement when they see Sam's car. A man is sitting on a bench. When Sam and his dad zip past him, his mouth drops open. He can't believe how slick and stunning the car is. *"Bellissimo!"* He shouts. This means "beautiful" in Italian.

"This is the most powerful car I have ever driven," Sam's dad says. He revs the engine.

PARTS OF A
FERRARI F8 TRIBUTO

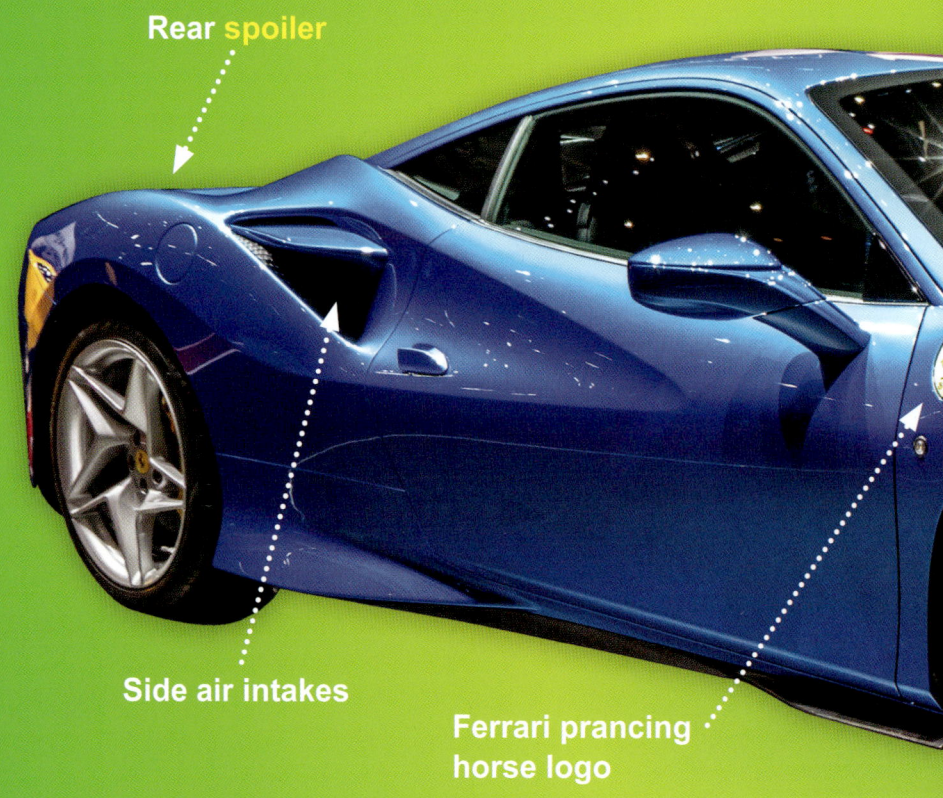

Rear spoiler

Side air intakes

Ferrari prancing horse logo

The F8 Tributo has 720 horsepower. Horsepower measures how powerful a car's engine is. The F8 Tributo has six times more horsepower than the average car in America.

FUN FACT

Why "Tributo"? Ferrari says this car's engine is a tribute to one of its earliest cars.

S-duct

Front air intakes

The Italian road winds and twists. Sam's dad handles each turn easily.

"This car has great turning power," he says. He has to shout. The engine is so loud!

The street suddenly becomes rough. Sam's dad presses the car's "Bumpy Road" button. As you go over some pot holes, the car remains smooth.

"Is there anything this car can't do?" Sam asks his dad. His dad smiles. "How about we hit Ferrari's test track and find out?"

"Let's go!" Sam responds.

THE PRANCING HORSE EMBLEM

The Ferrari emblem is a black prancing horse on a yellow background. The logo also has the letters S. and F. The S. stands for Scuderia. Scuderia is an Italian word for horse stable. It was adopted by racing teams. The F. stands for Ferrari. The horse was a symbol used by a famous Italian fighter pilot. It was painted on the pilot's plane during World War I.

Inside and out, the Tributo is beautiful. The seats are all leather. The car below is shown in the classic Ferrari red.

Chapter 2
Built For Speed

Enzo Ferrari was ten years old in 1908. He lived in Italy. One day, his dad took Enzo and his brother to their first road race. Cars were still pretty new at that time. Enzo loved it. He decided he wanted to race cars. When Enzo was 21, his dream came true. He became a race car driver. He raced for several years for another car company. He also helped build their cars. Soon he decided to start designing and building his own cars. In 1947, Enzo founded the Ferrari company.

Enzo Ferrari is on the far left. His father Dino is in the center. Together, they helped create the famous Ferrari engines.

Classic Ferraris on display at the company factory in Italy.

THE FIRST FERRARI

The first car made by Enzo was a Ferrari 125 S in 1947. It won six of its 14 races that year. It had a V12 engine and could reach speeds of 130 miles per hour (209 kph). It is worth over eight million dollars today.

Ferrari's cars began racing all over the world. They won again and again. Soon Ferrari wasn't just building race cars. He also made cars for people to drive on roads. Ferrari ran his company until he died in 1988. During that time, Ferrari cars won more than 5,000 races!

WHERE THE FERRARI F8 TRIBUTO IS MADE

Ferrari's name and company live on. Ferraris are still considered by many as the most beautiful road cars today.

Famous Ferraris include the Enzo, the Spider, and the Testarossa. You might also see the GT and the Dino.

Ferrari keeps making amazing new cars. The F8 Tributo was first sold in 2020. It was a great addition to a long tradition!

FUN FACT
The letters GT stand for Gran Turismo. These are Italian words for "grand tour."

Ferrari also makes cars that take part in the Formula 1 racing series.

Chapter 3
The F8 Tributo's Best Features

Sam can't believe his luck! His dad is taking him to the Fiorano Circuit in Italy. This is Ferrari's test track. Drivers can drive as fast as they want away from other cars. Sam's dad writes car reviews. He has tried the F8 Tributo on village roads. Now Sam and his dad get to push the car to its limits on the track.

Before Sam climbs into the car, his dad shows him the car's engine. It's unlike any one he has seen before. It is in the back of the car, not the front!

Ferrari is so proud of this engine, it is under glass. The Tributo's engine is a twin-turbocharged 3.9-liter **V8**. The Tributo can reach speeds of 60 miles per hour in 2.9 seconds! This makes it one of the fastest **accelerating** cars of all time!

FUN FACT
Car fans often compare a model's time to reach 60 miles per hour.

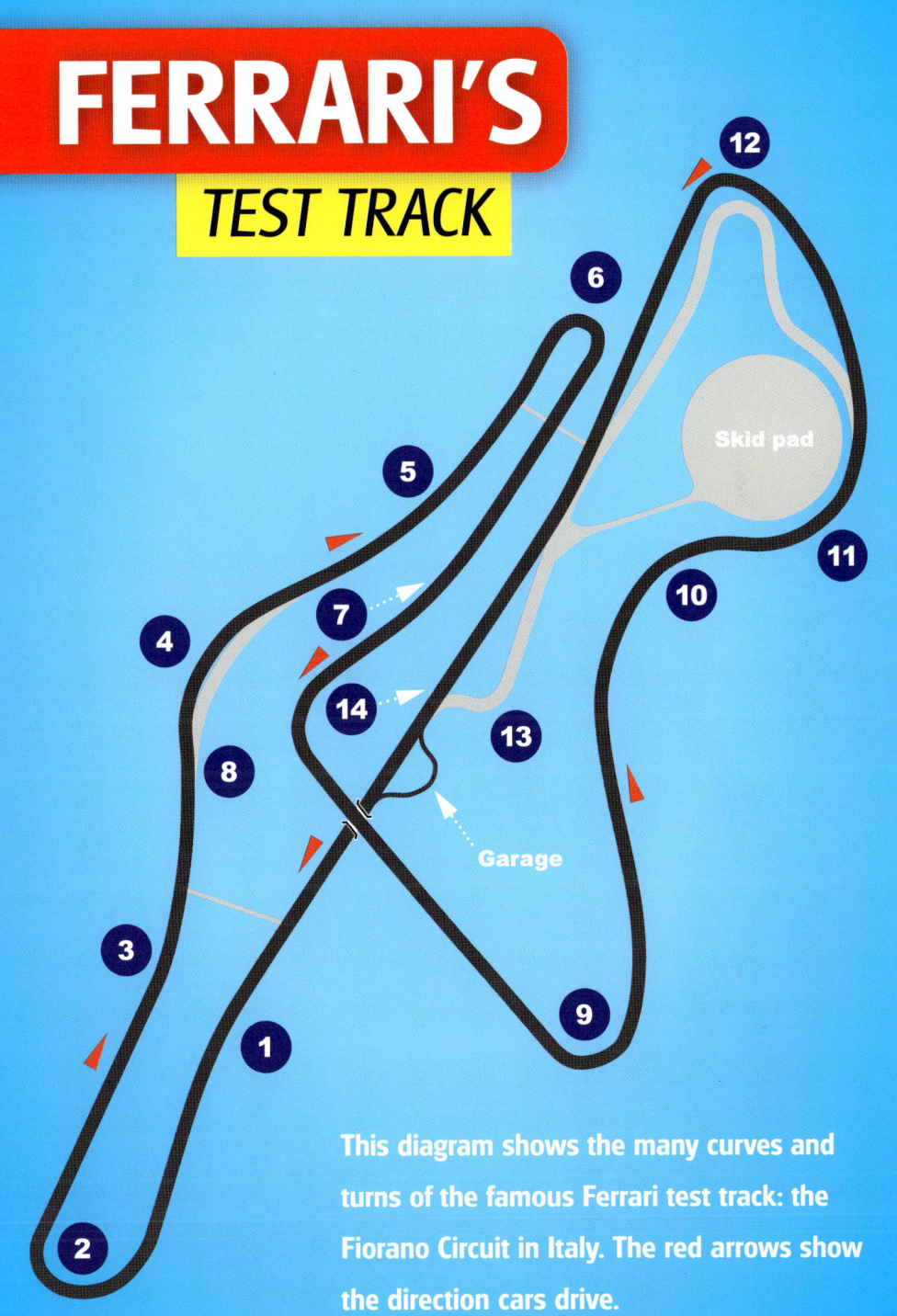

FERRARI'S
TEST TRACK

This diagram shows the many curves and turns of the famous Ferrari test track: the Fiorano Circuit in Italy. The red arrows show the direction cars drive.

Sam's dad checks that his seatbelt is on. He is ready. His foot hits the gas pedal. The car takes off. Sam can feel and hear the engine's power. His dad hugs the corner on a turn. The car is still going fast! On the **dashboard** in front of Sam is a seven-inch touchscreen. The digital numbers climb toward 100. The car accelerates after every turn. Sam has never gone this fast before!

One reason the car goes so fast is because it doesn't have a lot of air resistance.

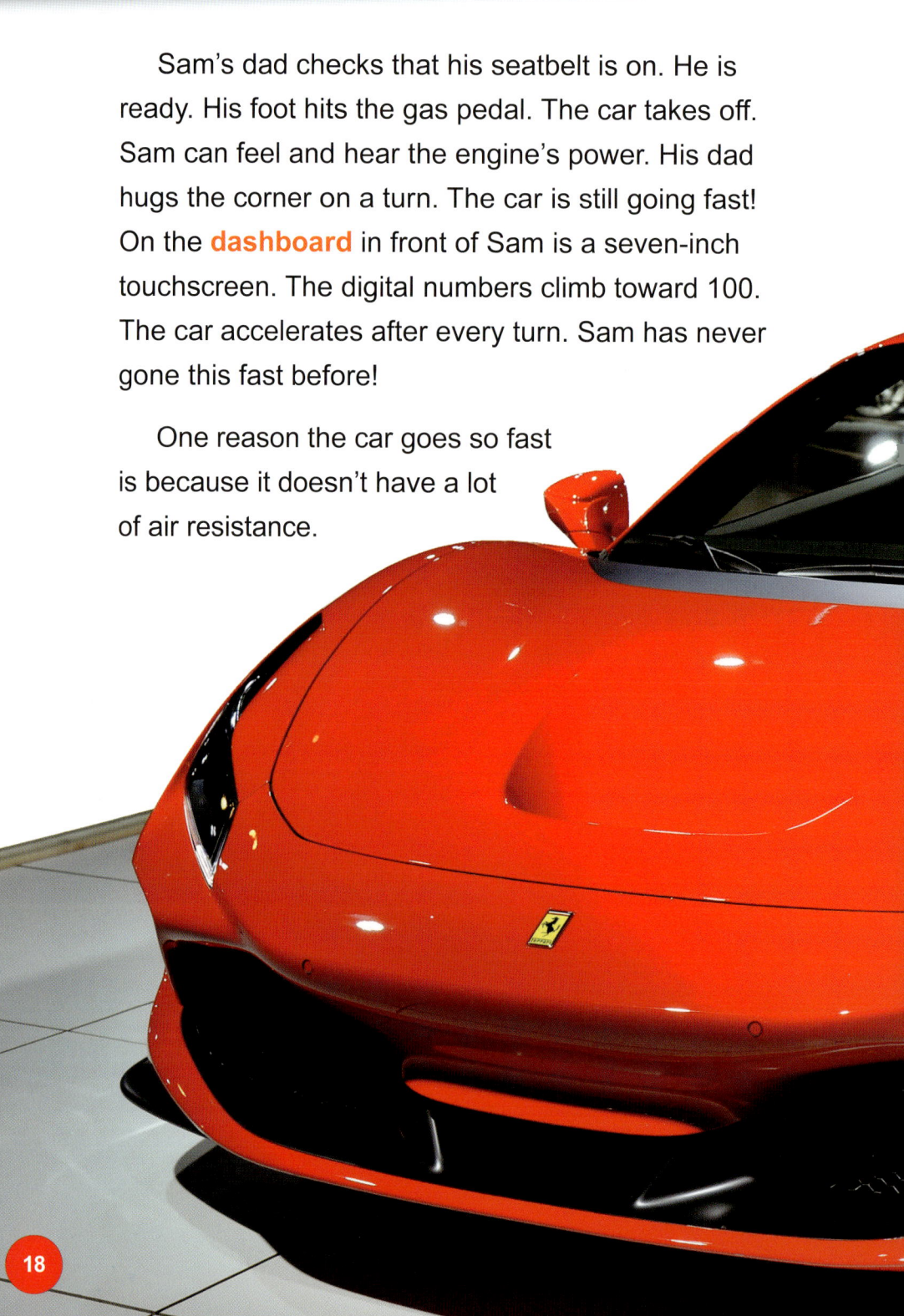

Air resistance slows down cars. The F8 Tributo lets air go through the front bumper. It comes out on top of the nose of the car. The air passes through the S-duct. This makes the F8 really **aerodynamic**! The spoiler at the back of the car helps do the same thing.

The Tributo is designed to have air flow very smoothly over its body.

THE FERRARI F8 TRIBUTO
IN DETAIL

Height: 3 feet, 11.5 inches (1.2 m)

Width: 6 feet, 5 inches (1.9 meters)

COST: $276,000 (United States)

LENGTH: 15 feet, 2 inches (4.6 m)

WEIGHT: 3,163 pounds (1,435 kg)

TOP SPEED: 211 miles per hour (339 kph)

TIME FROM 0-60 MPH: 2.9 seconds

Drivers like the small size of the Tributo's steering wheel.

The steering wheel is smaller than most other supercars. This makes it easier to handle. Sam's dad also tells him the car is super safe to drive. Computer controls help keep the car stable, even while it is going fast. It makes it seem like the car is gripping the road.

After a few laps, it was time to stop. His dad had to go write his review. Sam is left with only one question. "When can we drive the Ferrari F8 Tributo again?"

Chapter 4
The Tributo's Future

Ferrari cars get better and faster with each generation. This is especially true of the F8 Tributo. Some argue it is the finest Ferrari ever built. This car looks fast even when parked. It has a 568 pound-to-feet **torque**. This means this engine has a crazy amount of oomph. It's powerful, but it also uses its power wisely.

The Tributo also has a "Race" setting. This means less expert drivers can easily maintain control while twisting and turning along roads. The Tributo really triumphs over other racing cars.

THE LONNNG RACE

The 24 Hours of Le Mans is the world's oldest active **endurance** race. It is held in Le Mans, France over a period of 24 hours. Teams usually use three drivers. They cover distances over 3,000 miles (5,000 km). A Ferrari car has won this race nine times. Only Porsche (19) and Audi (13) have won it more.

FUN FACT
Some materials used for the Tributo's engine are also used for rocket engines!

The Tributo is named for a celebration, or tribute. It honors Ferrari's V8 history. It is a celebration of the amazing twin-turbo engine. The V8 has won no less than four Engine of the Year Awards over the past 20 years. The F8 Tributo looked to the past for inspiration. But it looked to the future and made everything better.

This car has 50 more horsepower than ever before. It is also 90 pounds lighter. The steering is what really makes this a car of the future though. It handles like a motorcycle around the tightest of turns.

Best of all, the sound of the F8 is like no other car—ever. The engine sings instead of growls. Cruising in a Tributo is like listening to an orchestra . . . while driving at the speed of light.

FUN FACT

The F8 Tributo comes in eight different colors: silver, black, white, yellow, blue, orange, gray, and racing red.

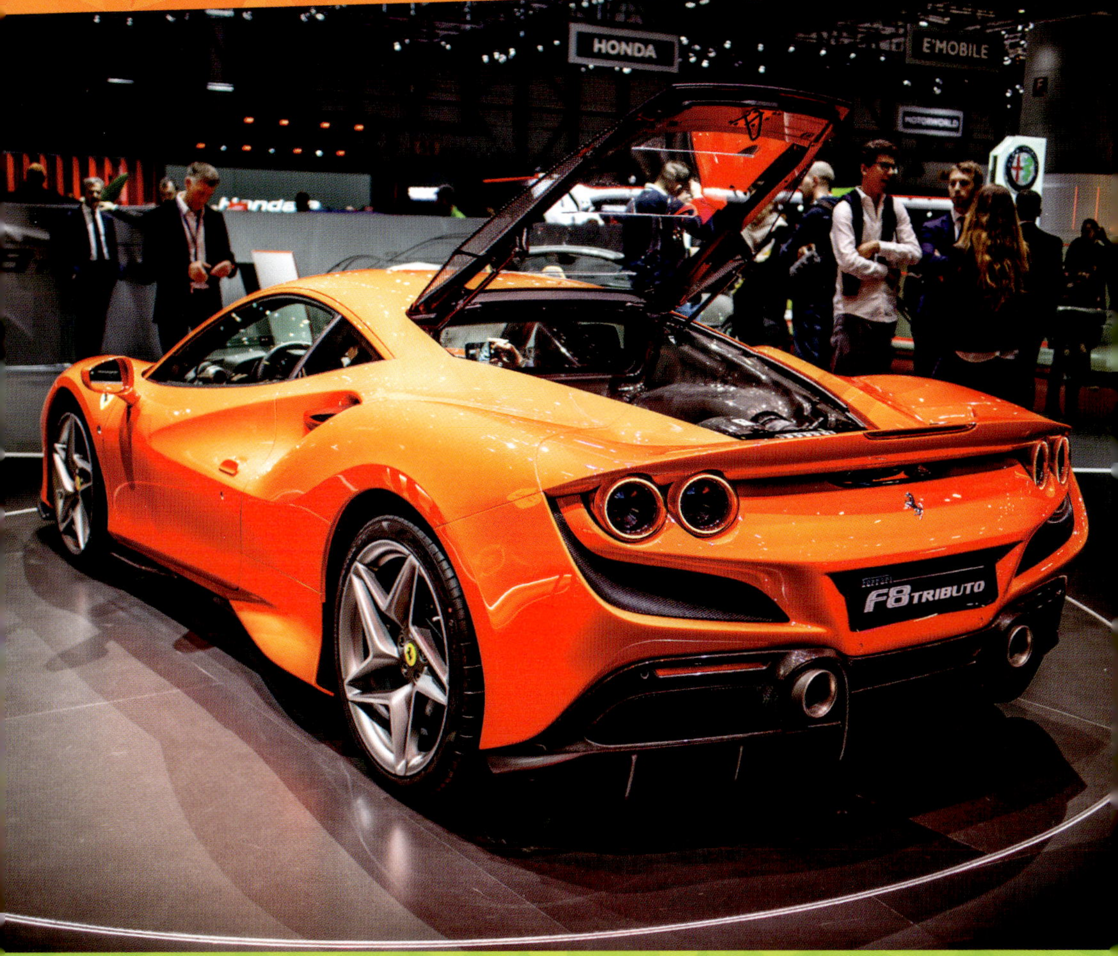
HONDA
E'MOBILE
F8 TRIBUTO

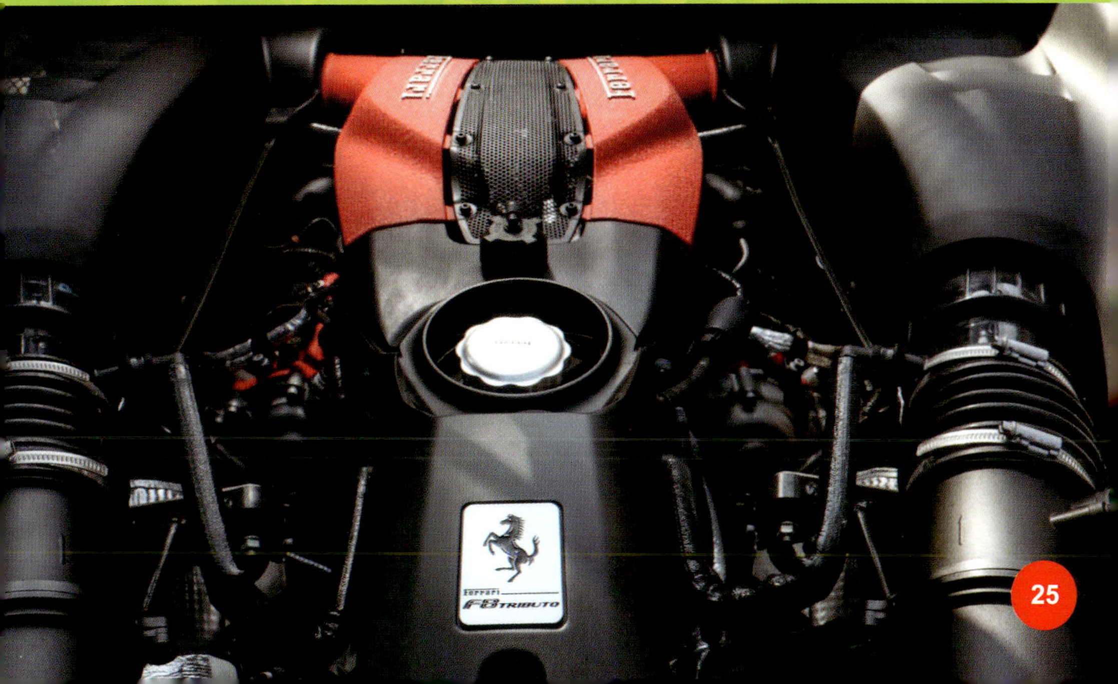
FERRARI
FERRARI
F8 TRIBUTO

The Tributo may be Ferrari's last car with a V8 engine. Future cars will probably be hybrid. This means the car will run on gas and electricity. It won't be easy to build something faster and more stylish than the Tributo. Still, faster and more stylish is Ferrari's business. *Ciao!* Ferrari. Until next time.

FUN FACT
Why hybrid cars? Electric engines are much better for the environment.

The Tributo also comes as a convertible. Feel the air in your hair!

BEYOND THE BOOK

After reading the book, it's time to think about what you learned. Try the following exercises to jumpstart your ideas.

RESEARCH

FIND OUT MORE. Where would you go to find out more about your favorite cars? Find out what company makes the car and locate its website. What information do the companies provide? What other sources of car information can you find?

CREATE

GET ARTISTIC. Cars start with creative artists and designers. Time for you to take a shot! Get art materials and create a great, new car. Will you make it a sports car? A sedan? A race car? What colors will you paint it? What features can you give it? Let your imagination go for a spin!

DISCOVER

DIG DEEPER. Look more into Ferrari's racing history. What other big races or series has it won? How do the cars on the race track inspire cars you see on the street? What are the most famous Ferrari cars and drivers among pro racing?

GROW

GO TO A CAR SHOW. Car shows are a great way to see lots of cool cars up-close. Check your local events calendar, or ask at a car dealer for upcoming events. You can find shows of old cars and new cars, sports cars and classic cars. Go to a show and find a new favorite car to love!

RESEARCH NINJA

Visit www.ninjaresearcher.com/2626 to learn how to take your research skills and book report writing to the next level!

RESEARCH

DIGITAL LITERACY TOOLS

SEARCH LIKE A PRO
Learn about how to use search engines to find useful websites.

FACT OR FAKE?
Discover how you can tell a trusted website from an untrustworthy resource.

TEXT DETECTIVE
Explore how to zero in on the information you need most.

SHOW YOUR WORK
Research responsibly— learn how to cite sources.

WRITE

GET TO THE POINT
Learn how to express your main ideas.

PLAN OF ATTACK
Learn prewriting exercises and create an outline.

DOWNLOADABLE REPORT FORMS

Further Resources

BOOKS

Publications International. *Built for Speed: The World's Fastest Road Cars.* Morton Grove, IL: Publications International, 2019.

Roach, Martin, Neil Waterman, and John Morrison. *The Science of Supercars: The Technology that Powers the Greatest Cars in the World.* Ontario, Canada: Firefly Books, 2018.

Storm, Marysa. *Supercars (Wild Rides)* Mankato, MN: Black Rabbit Books, 2020.

WEBSITES

Factsurfer.com gives you a safe, fun way to find more information.

1. Go to www.factsurfer.com.
2. Enter "Ferrari F8 Tributo" into the search box and click 🔍
3. Select your book cover to see a list of related websites.

Glossary

accelerate: to go faster. The Ferrari F8 Tributo can accelerate from 0-60 miles per hour (97 kph) in 2.9 seconds.

aerodynamic: a car design that reduces drag. The F8 Tributo is one of the most aerodynamic cars ever.

Ciao!: "goodbye" in Italian (pronounced "chow!").

dashboard: a panel located in front of the driver and passenger of a car. The F8 Tributo's dashboard tells the passenger how fast the car is going.

emblem: a car's logo that represents the car's brand. Ferrari's emblem is a black prancing horse.

endurance: ability to last for a long time.

S-duct: an inlet on the front of the car that helps reduce the airflow over the car. Thanks to the F8 Tributo's S-duct, the car goes faster than ever.

spoiler: the wing on the back of a car that changes the airflow. The Ferrari F8 Tributo has a spoiler that helps make it more aerodynamic.

torque: the turning power of a car. The torque on the Ferrari F8 was crazy fast.

V8: an engine that has 8 cylinders in the shape of a V. The Ferrari F8 Tributo has one of the best V8 engines there is.

Index

24 Hours of Le Mans, 22
125 S, 12
Audi, 22
Dino (car), 14
engine, 5, 6, 7, 8, 10, 12, 16, 18, 22, 23, 24, 26
Enzo (car), 14
Ferrari, Dino, 10
Ferrari, Enzo, 10, 12
Ferrari logo, 6, 8
Fiorano Circuit, 16, 17
Gran Turismo, 14
handling, 21, 24
hybrid car, 26
interior, 9
Italy, 4, 10, 11, 13, 16, 17
Porsche, 22
Spider, 14
Testarossa, 14

PHOTO CREDITS

The images in this book are reproduced through the courtesy of: Courtesy Ferrari: 9. Shutterstock: Grzegorz Czapski 4, 25; VanderWolf Images 6; Lawrence Carmichael 9T; Dan74 12; Cristiano Barni 14; Tanase Sorin 18; Rasta 777 20; Kullapong Parcherat 21; Terry Leung 26; auto-data-net 27. Wikimedia: 10; Will Pittinger 17; Matti Blume 23; 24.
Cover: Lawrence Carmichael/Shutterstock (car); YIUCHEUNG/Shutterstock (background, top); zhao jiankang/Shutterstock (background, bottom).

About the Author

Ellen Labrecque has written many non-fiction books for kids. Previously, she was a senior editor at Sports Illustrated Kids where she enjoyed following the race car scene. She lives outside of Philadelphia with her husband, two kids, and her dog Oscar—who also happens to be the best writing partner ever.